I0454599

Yaya and Wormie's Space Adventure

An Original by Ahllah Azwar,
"Love doesn't have a predefined face nor a
particular race."

Copyright 10/01/2023

To:

From:

About this book:

A Fully illustrated children's educational book about the planets and the cosmos, filled with facts about the universe and how it works.

Making learning via a story about how two most unlikely characters can become friends. The story revolves around two enemies; A Worm and a Chicken. They go on a space adventure together to learn about the planets in the universe. This adventure leads them to the greatest journey of friendship.

(Building rockets, meeting aliens, visiting other planets and learning life lessons).

Filled with humor, adventure, and lessons learned. **C**omes with a separate Interactive activity book.

Table of Contents

PART 1:

THE DEAL

Yaya the chicken was awakened by Mr. Roo-Roo with a loud "cock-a-doo-dle-doo" right in her ear! Mr. Roo-Roo the rooster did that every morning, waking her up way too early. Yaya looked outside and it was still nighttime.

A rooster is supposed to crow when the sun wakes up, but Mr. Roo-Roo always crows way before the sun rises; never allowing her a full night's sleep. She wished she could sleep in just a little longer. Knowing that will never happen with Mr. Roo-Roo around, she decided to go get breakfast instead.

Jumping down from her chicken roost, Yaya ran straight to where Mr. Henry always leaves her breakfast, but when she got there, she found only last night's leftovers. Disappointed in the food selection, Yaya left the chicken coup to find her own breakfast.

Yaya looked outside and it was still a little dark. Feeling tired, and hungry made her extremely grumpy. She was thinking of how Mr. Roo-Roo is so selfish to wake everyone up so early. 'He has no consideration for others' feelings'. She thought as she stared out at the meadow just outside her coupe,

Yaya's eyes spotted something moving. She stepped closer to realize it was a huge juicy worm. She smiled thinking out loud "My day just started to get much better."

As Yaya got closer she saw that half the worm was still in the ground and it was quickly trying to get back under. Knowing she only had a limited time span to catch her yummy breakfast; she ran quickly to pluck the worm out of the ground before it could get away.

Grabbing the worm with her beak, she pulled as hard as she could. Suddenly, she heard a scream coming from the worm; "What are you doing?" Shocked to hear her breakfast questioning her, Yaya took a step back and said: "I'm trying to have my breakfast!"

Nervously Wormie answered Yaya saying: "Now hold on a bit, I have had a really bad day and you having me for breakfast doesn't help make my day any better."

Yaya decided to entertain her breakfast because she was curious about how worms can have a bad day. She thought about how bad her day had been so far, and curiously asked why he was having a bad day.

Wormie replied: "Every morning Mr. Roo-Roo wakes me up super early. I never get a chance to sleep in, but today Mr. Roo-Roo woke me up and it's STILL NIGHT TIME! And now you're trying to eat me for breakfast before the sun has even risen." Yaya thought about what he just said and remembered how Mr. Roo-Roo wakes her up all the time too.

Yaya resonated with Wormie's hardships and started to feel bad, but then quickly thought, 'this worm is trying to trick me, so I don't eat him.' Yaya made sure he knew she was not fooled so easily revealing "Worms can't have bad days! Worms are made for eating and are on the bottom of the food chain and have no rights to their feelings! Therefore they can't have a bad day like a chicken can, so too bad, I'm having you for breakfast!"

Wormie: "Fine, then Let's make a deal."

Yaya: "What kind of deal?"

Wormie nervously responded: "I will let you eat me and promise not to run away if you could just spare my life for one day."

"Why would I do that?" Yaya amusingly asked, and then directly looked into his eyes and said: "YOU'RE NOT IMPORTANT!"

Wormie felt hurt and wished that he couldbe as important as chicken, and not just some object to be eaten and pooped out. But Wormie kept his feelings to himself because he was afraid that she would just laugh at him or even worse, eat him right away.

Instead of sharing his feelings, Wormie said:"I have always been underground and I never had a chance to explore. All I ask is that you spare my life until I see those pretty lights dancing in the skies up close and that big round floating ball that changes its shape but comes out almost every night. "

Yaya: "YOU WANT TO GO ON A
SPACE ADVENTURE!"

Wormie: "uh... Yes"

Yaya was confused and not sure what to think of all this. She never had her 'breakfast' try to make a deal with her. She didn't even know worms had dreams and feelings like chickens do. She looked up into the sky and saw the pretty lights fading as the sun slowly rose. Thinking to herself, a closer look would be nice and some time away from the loud Mr. Roo Roo would be worth delaying breakfast for a day. But what would the other chickens think if they saw her hanging out with a worm, she wondered.

They would make fun of her for sure, she thought. Thinking about how embarrassing it would be to go on a space adventure with a worm named Wormie, She could let him go without her, but then quickly dismissed this thought. Yaya realized she wanted to go too.

Yaya thought about it a little longer and finally said: "Okay, I will let you go on a space adventure but I'm going with you! Then after our adventure I will eat you for breakfast and if anyone sees us together, you're just my breakfast! I don't want anyone to know if I hang out with you or that you're my friend. That would be embarrassing."

Wormie didn't think Yaya was very nice; her words hurt him and he didn't want her to join him but he was happy that he bought himself some time to figure out an escape plan.

Wormie replied: **"Deal."**

Yaya and Wormie needed to build a rocket ship, and not too far from the chicken coop there was a metal scrap yard. A perfect place to find all the materials and parts required for a rocket. Yaya and Wormie decided to grab some supplies from the chicken coop before going to the scrap yard to build their rocket.

When they entered the chicken coop all the chickens ran over to Wormie wanting to eat him. Yaya screamed out loud: "Wormie is my breakfast and I'm saving him for later, so no one is allowed to eat him but me!"

The other chickens all started laughing at Yaya for playing with her breakfast. They called her names and told her she had no friends except for a worm.

Yaya lost her temper and angrily said: "He is not my friend! I am just saving him for when I'm hungry!" Yaya then looked over at Wormie and saw he looked scared. She felt bad for him, bullies she thought.

Yaya: "Common Wormie, let's go build our Rocket. It's starting to smell bad in here."

The other chickens started laughing even harder and louder as they both walked out.

PART 2:

Yaya And Wormie

BUILD A ROCKET!

When Yaya and Wormie got to the scrap yard they started to look around for metal pieces they could use to build their rocket. They knew that they would need to compromise on some of their own ideas for how the rocket would look, if they wanted to finish before the sun went down.

They spent the entire day working together building their rocket. It required both Wormie and Yaya to compromise on some of their own ideas for how the rocket would look, if they wanted to finish before the sun went down.

It was almost dark outside when both Yaya and Wormie stopped and looked at their Rocketship and then turned to look at one another. Neither said a word, but both knew what the other was thinking. 'We built a Rocketship together and it was fun!'

"UP UP AND AWAY!!" Wormie and Yaya both yelled out at the same time and couldn't help but laugh out loud as they rocketed out to space.

Yaya and Wormie were so surprised at how beautiful it was in outer space. They saw planets and the stars shining so brightly, it was magnificent. As they got closer to the floating ball they had seen from earth, both agreed to land there first.

When they landed on the floating ball, they got out of their rocket to explore.
Yaya heard her belly rumble in hunger. She still hadn't eaten anything and started to feel super hungry. She looked at Wormie thinking 'just one bite.

Wormie caught Yaya looking at him and knew what she was thinking and quickly said: "Don't forget our deal!"
Yaya smiled to herself; secretly knowing that the truth was that she would rather be hungry with Wormie verses full without him.

PART 3:

Yaya And Wormie

Meet

Mr. Comosho

(MOON ALIEN)

Just then an little green alien popped up and said: "Well hello earthlings, welcome to the moon. I am Mr. Comosho, who might you guys be?"

Yaya and Wormie both replied with their names, and then told Mr. Comosho about their planned adventure and how they really wanted to see what this place because it looks like a big floating ball from where they come from. "That's a fantastic idea, I always love a good adventure, would you like me to be your guide?" asked Mr. Comosho.

Yaya and Wormie were very pleased with Mr. Comosho's offer, and both quickly accepted.

Mr. Comosho laughingly said: "A floating ball uh? We will take my spaceship and I can show you around the Cosmos."

Mr. Comosho led the way towards his spaceship and decided to start the tour by telling Yaya and Wormie about the moon that they had called a floating ball. "This is the place where I live. It's called the moon.

There are several moons in our cosmos, but this moon is the one you see on earth, the one you call the 'floating ball'. The floating ball hehehe" He chuckled.

"What is the Moon?" Asked Wormie

Mr. Comosho responded: "A moon is a celestial body that orbits a planet.   This moon goes around the earth and because moons are smaller than planets, they are the ones that orbit and are held in their orbit by the planet's gravitational pull. This is the Earth's only natural moon satellite. and this moon is the fifth largest moon in the solar system."

Yaya and Wormie were so confused. Wormie was the first to ask: "What does celestial body mean Mr. Comosho?"

Following Wormie's lead, Yaya also asked a question: "What does it mean to orbit?"

Mr. Comosho Laughed and said: "One question at a time, I will answer Wormie's question first.

A celestial body is anything that is in space and is there naturally. For example; I am considered a celestial body, because I am not man made. I was created by the one and only creator of the universe to exist on the moon and in space. You guys were created naturally to live on earth and are considered aliens to us here on the moon.

"**I**s our rocket a celestial body?" Wormie asked.

"**W**ell number one: your rocket was not born as a Rocketship, that means it's not a natural creation that was born or formed like that without intervention. It had to be formed by you into a rocketship by you so therefore it's not a celestial body. Thw celestial body is only what our creator creates directly without any intervention. Your rocket ship is considered an alien spacecraft here on the moon because it was not made on the moon. So it's strange to us here on the moon. When something is strange or doesn't belong to a place we call that alien to us because it is different from what we are used to." Mr. Comosho replied.

Wormie had to think about it for a bit and asked him a follow up question just to make sure he understood.

"**M**r. Comosho, I am naturally made by our creator, and I am in space right now. Does that mean I am a celestial body right now?"

Mr. Comosho laughed and said: "That is a very good and tricky question. You see, a celestial body is always located outside the earth's atmosphere and is naturally created. We all become celestial bodies in the end because our soul is celestial, but your physical body is not celestial and made to be on earth. So, in answer to your question, you are not a celestial body right now because you have a physical body suit on. However, when your soul leaves your bodysuit then your soul is considered a celestial body. So as long as your soul is attached to your physical body you are not considered a celestial body. Right now, you are a visitor just like your Rocketship, and we call you an earthling", replied.Mr. Comosho.

"**O**h" Wormie said and started to understand some more but still had one final question to make sure he understood completely. "So, everything with a soul eventually becomes a celestial body when they leave their physical body, is that correct?"

"**Y**es Wormie, our souls are eternal, and our physical world is just a temporary experience. We are all part of the sparkling lights you see in the night sky. That is why none of us is really that different, it's just our body suits that make us look different and hence trick the ones that don't know better ", smiled Mr. Comosho.

Yaya cut in and said: "Is it my turn now?"

Mr. Comosho said: "Yes Yaya, it's your turn. Now what was your question again, hmmm ... oh yes , you want to know what it means to orbit. You see an orbit is just a path that goes around bigger objects in a circle,repeatedly. So. an object that is in orbit is an object that is moving in a big circle around another object. For example, the moon is smaller than the earth, so it goes around the earth in a circle. Does that now make sense?"

Yaya smiled and said: "So an orbit is just going around something bigger over and over again in a big circle?"

"Yes, basically that's all it is", said Mr. Comosho.

Mr. Comosho started to give them both some more fun facts about the moon.

"Our Moon plays an important role on the earth. The moon helps with the way you tell time, and the cycles of life. It's looked upon by many as a tool for guidance. The moon can affect the ocean tides and even your emotions. But not all moons look the same, some moons have mountains, and valleys, while others are covered in ice. Did you know that the earth has one moon while the planet Jupiter has 92, and astronomers still haven't found all the mysteries of space."

"Jupiter? What's That", asked Yaya.

Mr. Comosho laughed and said: "It's a planet, just like the earth is a planet, " as he pointed to Jupiter visible in front of them.

Changing the tone of his voice, Mr. Comosho said: "So are you guys ready to go for a ride in my spaceship?"

Yaya and Wormie replied enthusiastically with "YES SIR!"

"**G**ood, common then, I'll show you what earth and other planets look like from space." Said Mr. Comosho as they jumped into the spaceship and rocketed out into space.

PART 4:

Yaya, Wormie And Mr. Comosho

EXPLORE THE GALAXY

ALIEN-SPACESHIP

Yaya and Wormie were looking out the spaceship window while listening to Mr. Comosho tell them about the solar system, and how the planets all orbit in space.

"Look at that," Yaya said excitedly.
"'It's so big and bright!" Added Wormie.

Mr. Comosho chimed in: "That's the planet Jupiter. Did you know that the planet Jupiter is the fifth planet from the Sun. Jupiter can be easily seen from earth in the night sky."

"I think I've seen it before," Yaya excitedly said.

"Yes, you have, that's the same one I showed you from the moon, and you both have probably seen it several times from earth and now this is the view from space. Did you know that the planet Jupiter has been known since ancient times. Its name comes from the Roman god Jupiter. It is usually the third brightest object in the night sky after the Moon and Venus. Some people say that Jupiter is the luckiest planet," replied Mr. Comosho.

Yaya's tummy rumbled when she looked over at Wormie. She started imagining how tasty he would be, but then quickly stopped her thoughts. She was having so much fun with him and knew she would protect him from both her and others if ever required. He was her friend.

"What are planets?" asked Wormie.

"Planets are round objects that orbit stars and have their own gravity. Our solar system several planets, including Earth.   New planets are continously being descovered.   Would you guys like to see some of the main planets" Asked Mr. Comosho.

Yaya and Wormie replied excitedly with a "YES PLEASE!" Mr. Comosho showed Yaya and Wormie 8 planets: Mercury, Venus, Earth, Mars, Staturn, Uranus, Neptune, Jupitor and a view of Earth from outer space.

The Planet Mercury

The Planet Venus

The Planet Mars

The Planet Saturn

The Planet Uranus

The Planet Neptune

The Planet Jupiter

The Planet Earth

They both thanked Mr. Comosho and let him know that if he ever wanted to visit the planet Earth that they would be his tour guide next time.

Mr. Comosho was very happy to receive this invitation. He looked at them both and said: "I feel honored to have met you both. You earthlings are very special and have shown me that earthlings can love unconditionally and that friendships do not require a certain face or need to be a particular race. You see, I watched you sacrifice for a friend Yaya. I knew you were hungry; I heard your tummy rumble throughout our tour, and you could have easily eaten Wormie and taken the hunger pains away. After all, worms are favorite snacks for chickens. Instead, you made a choice to not eat Wormie and deal with the pangs of hunger. You showed me that earthlings are capable of sacrifices for the sake of the another. Thank you for showing me what earthlings are capable of... Again, I am honored to have met you both. I hope we meet again soon."

Mr. Comosho waved goodbye as Yaya and Wormie rocketed back to earth.

PART 5:

WHAT'S THAT

SMELL?

When Yaya and Wormie arrived home they were greeted by a flock of chickens.

They stared at Yaya and Wormie. Yaya told Wormie to stay close. She didn't know what these chickens were planning to do.

Wormie: "Thank you Yaya, I'm super scared."

Yaya: "Don't be scared Wormie, I won't let anything happen to you, I promise. Hop on my back and hold tight in case we need to run."

Wormie hopped on Yaya's back and grabbed a bunch of her feathers and held on super tight.

"OW! Not that tight," Yaya screamed.

"Oh sorry, sorry," Wormie said, as he   loosened his grip.

"What was that?" Yaya shockingly asked.

Wormie embarrassed said: **"I'm so sorry, I fart when I get scared."**

Wormie kept farting nonstop on Yaya. Yaya was trying to hold on to her anger because she knew that he was in danger, and she didn't want to lose him. Yaya realized that she loved him and would protect him. He was her friend, and she didn't care what anyone else thought. Yaya gathered her courage and started walking towards a bunch of angry chickens.

Only a few feet away from the angry flock of chickens, Yaya felt something warm and wet hit her back. Horrified, Yaya whispered slowly: "What was that?"

Wormie was afraid to answer her. "What did you just do Wormie?!" knowing he had to respond after she asked again, Wormie said: "I'm sorry Yaya, I can't help it. I'm so scared and lost control"

Yaya: "Did you just poop on me?"

Wormie: "I'm sorry, I'm so sorry…"

Appalled, Yaya clenched her beak together and threatened in a low whisper: "I will eat you myself if you don't stop pooping and farting on me!”

Yaya stepped closer towards the angry chickens knowing there was no other way out unless she talked to them to see why they were so angry.

Freeda the hen was the first to talk:
"Yaya, we all thought you were joking when you said you were building a rocket and going ON A SPACE ADVENTURE with a WORM! When we realized that you meant it, we all had a meeting and expressed our thoughts and concerns about your decisions and actions.”

Freeda the hen continued to say; "We realize now that you needed a friend and that we have been too busy with our chicks and lives and failed to notice you felt lonely and needed us. We were not there for you and take full accountability."

Wormie continued to fart, but the other chickens couldn't hear anything due to all the noise. Yaya felt a wet warmth on her back and couldn't wait to get him off her back and take a bath. She wished Freeda would stop talking so she could leave.

Freeda confidently comtinued, "Yaya, we are chickens, and we eat worms for breakfast because they are lower in the food chain than us. It is not acceptable to be friends with such a lower class. So, we voted unanimously to assign Sujo the hen to be your new friend."

Before Freeda could finish, Olivia the mama hen screamed out: "Eat the worm so our chicks don't get confused. They need to see that worms are for breakfast and not made to be friends! EAT THE WORM YAYA!"

Yaya responded to Olivia with an angry, "NO, Wormie is my friend, he is funny, loving, smart and fun! I WILL NOT EAT MY BESTFRIEND!"

Wormie was touched to hear Yaya say all those lovely words about him, and for a moment forgot about being scared and stopped farting. He was so happy. However, It didn't take long before Wormie was snapped out of his happy thoughts by the angry chickens chanting: **"KILL THE WORM!"**

The Chickens were getting angrier as they violently chanted: "kill the worm!"

Yaya started getting scared. She didn't want Wormie to get hurt. She didnt care what they thought about how a chicken and worm cannot be friends because they are too different. She liked their differences, and she didn't see Wormie as being lower in status than her anymore. She saw him as her equal, and as a friend. She decided that she would protect him no matter what happened. Afterall, she would want someone to have her back if it was the other way around.

Other chickens started to chant in and got louder and closer. **"KILL THE WORM! KILL THE WORM! KILL THE WORM!"**

"WHAT'S THAT SMELL!"

After Oliva the hen horrifically yelled "WHAT'S THAT SMELL!" All the chickens stopped chanting and an uncomfortable silence proceeded.

They all looked at Wormie shocked. Wormie got so nervous which caused his farts to get stinkier and louder. So loud that even the chickens in the back were horrified. The chickens stared at him in horror, they were disgusted. Wormie wished they would start yelling again, anything was better than the silence.

Terrified. all Wormie could think to say was: "I'm sorry, I can't help it, I fart when I get scared."

PART 6:

Roo-Roo The ROOSTER AND

The New Law!

Then Roo-Roo showed up looking baffled as he turned his gaze directly at Yaya and said: "What is going on here? What is all this noise about?" I can hear you hens squawking all the way from the chicken coop!"

Roo-Roo's gaze then moved to Wormie. "What is this Yaya? Why is your breakfast on your back instead of in your mouth? And what is that awful smell?" Everyone quietly awaited Yaya's answer.

Meanwhile, Wormie was squeezing his butt cheeks in together, trying so hard to hold his farts in. He didn't want Roo-Roo to know, except he didn't know how much longer he could hold it in.

Yaya said: "First of all the worm on my back has a name. His name is Wormie, and he is my friend, not my breakfast. I will not eat him and will not allow anyone to hurt him either!

Roo-Roo stepped closer to Wormie as he suspiciously looked Wormie, then then back at Yaya, still confused and calmly asked Yaya again, "what is that smell?"

Wormie lost control and farted so loud, causing Mr. Roo-Roo to jump back startled.

Now I really messed up, thought Wormie, and started to apologize profusely.

"I'm sorry, I'm sorry, I was trying to hold it in but just then, when you looked at me I got scared and lost control... I fart when I get scared."

Roo-Roo the Rooster burst out laughing and said, "I wouldn't eat him either!"

Then all the other chickens joined him in laughter.

When Roo-Roo finally stopped laughing he said: "Yaya you and your friend need to go to the river and take a bath. Nice to meet you, Wormie. I'm glad Yaya has a new friend. Just please stop farting, there is no reason to be afraid.

So here is the deal, I will not allow any of the other chickens to hurt you as long as you stop fartingou are safe here." Roo-Roo couldn't help but laugh again.

Roo-Roo then turned to address the entire flock:

"Yaya is allowed to have friends and they don't have to be just chickens. No one will harm nor make fun of either of them, or they will deal with me!

One of the chickens screamed out: "Worms are for breakfast not for friends!"

Then another chicken screamed out: "Worms are dirty, they live in the dirt and eat dirt, they are beneath us", and others continued to yell out: "We cannot damage our reputation by allowing a lowly worm to sit with us and think they can be our equal!"

Roo-Roo was shocked to hear how prejudiced and negative his hens were. Angered, Roo-Roo yelled: "STOP!"

All the chicken went silent. Now I will not tolerate this type of abuse to any living creature. You are no better, nor any worse, we are different, and all living things have purpose. I will not permit anymore of this ignorance and abuse to any living creature. If I hear any further squawking regarding Wormie, you will no longer be a part of my flock and can go find another home. Does everyone understand me?" All the chicken went silent and slowly nodded in acceptance.

Yaya looked at Roo-Roo with different eyes than she had before. She had never seen this side of Roo-Roo before. He was caring and kind. She realized that maybe she didn't see his kindness because she wasn't kind. She thought like the other chickens before getting to know Wormie. She felt ashamed of how she was in the past and promised herself that she would never judge another again.

Roo-Roo looked at Yaya and Wormie and playfully said. "You two go on now, you're not smelling any better."

Yaya looked at Roo-Roo and gave him a smile of appreciation before She and Wormie headed to the river.

THE END

PART 2:

Mr. Comosho Visits Earth

(coming soon...)

www.ingramcontent.com/pod-product-compliance
Lightning Source LLC
Chambersburg PA
CBHW082143290526
45794CB00008B/3150